1 MONTH OF
FREE
READING

at

www.ForgottenBooks.com

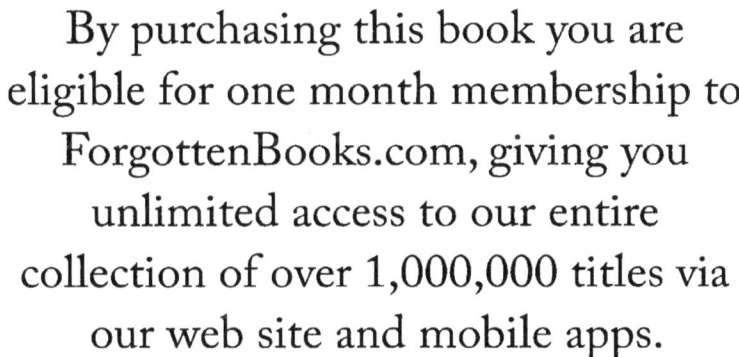

By purchasing this book you are
eligible for one month membership to
ForgottenBooks.com, giving you
unlimited access to our entire
collection of over 1,000,000 titles via
our web site and mobile apps.

To claim your free month visit:

www.forgottenbooks.com/free991073

ISBN 978-0-332-68720-9
PIBN 10991073

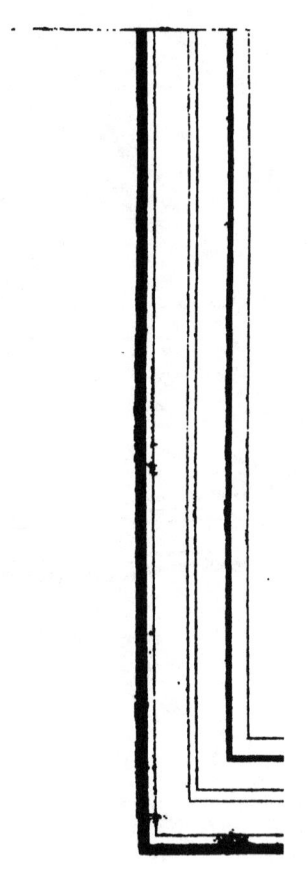

Map
OF THE
Vilage of
A-TITLAN.

Drawn by Cha: C. Smith.

1851.

Scale 800 Ft. to one Inch.

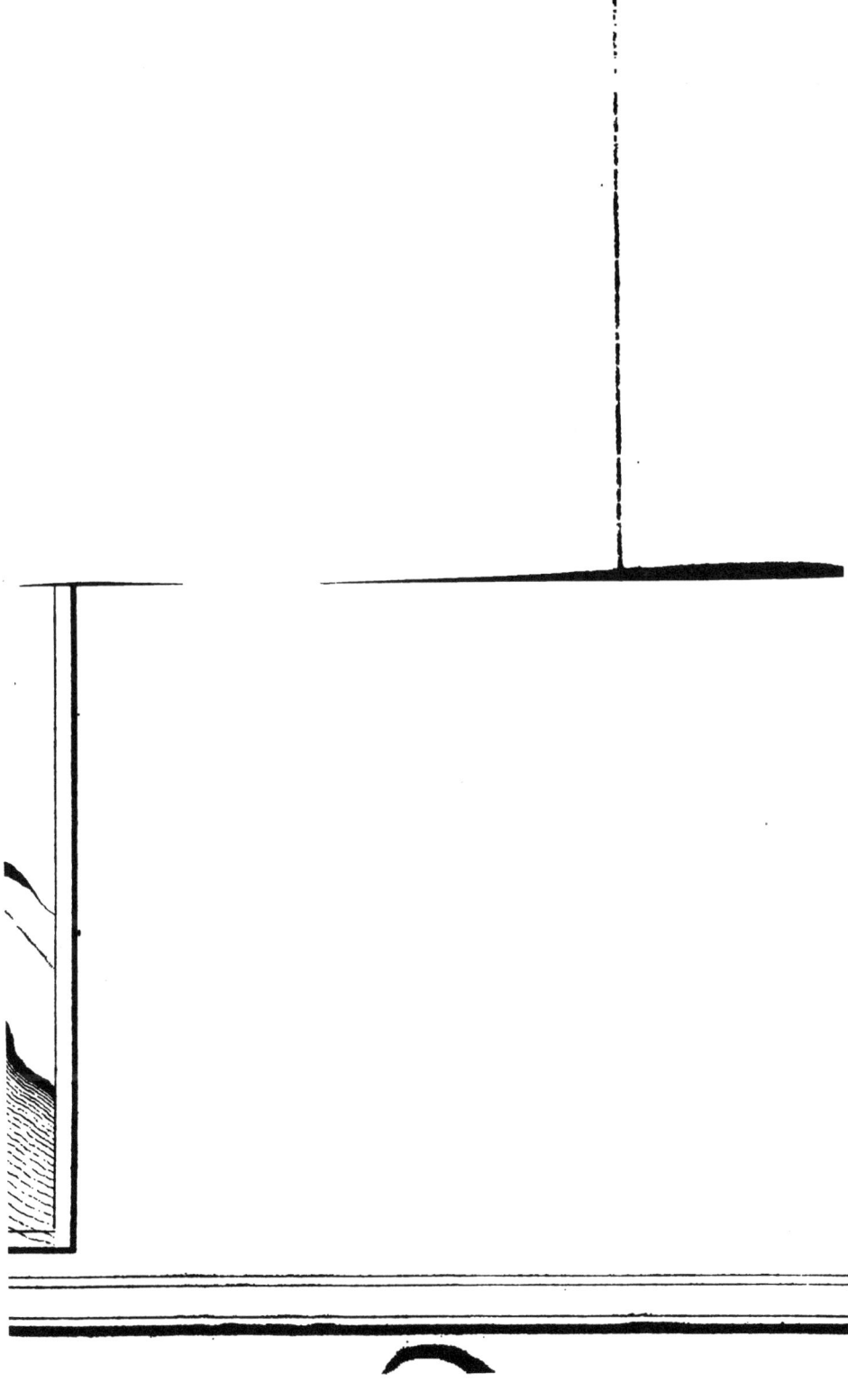

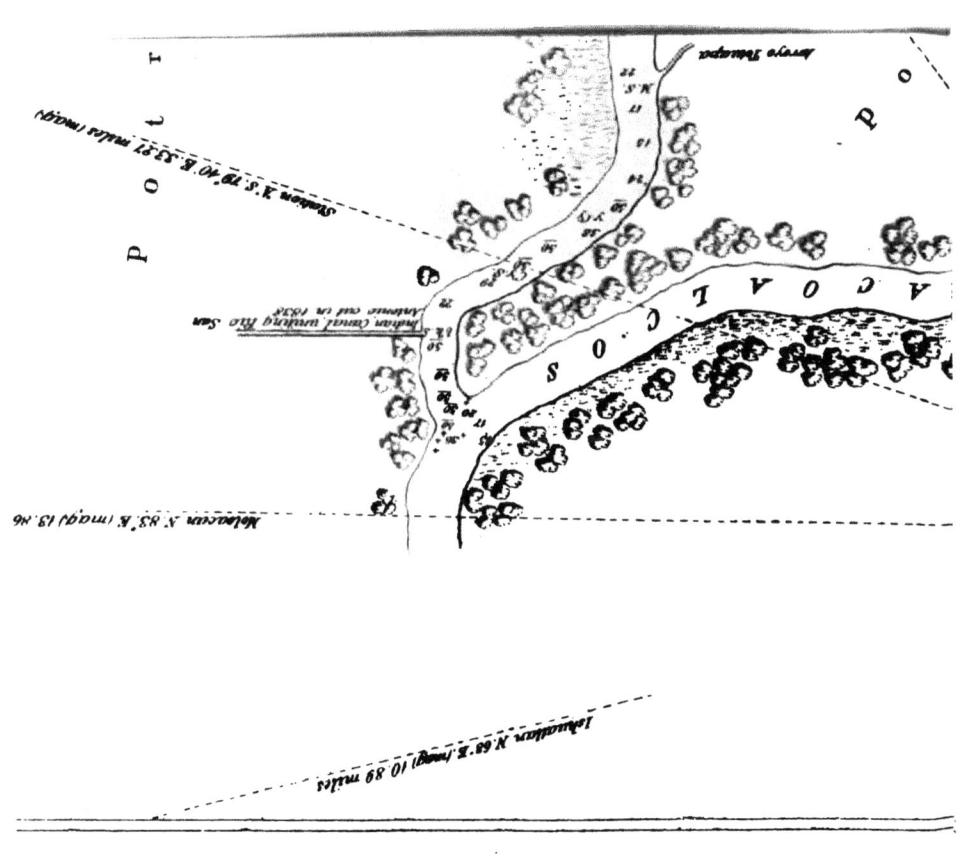

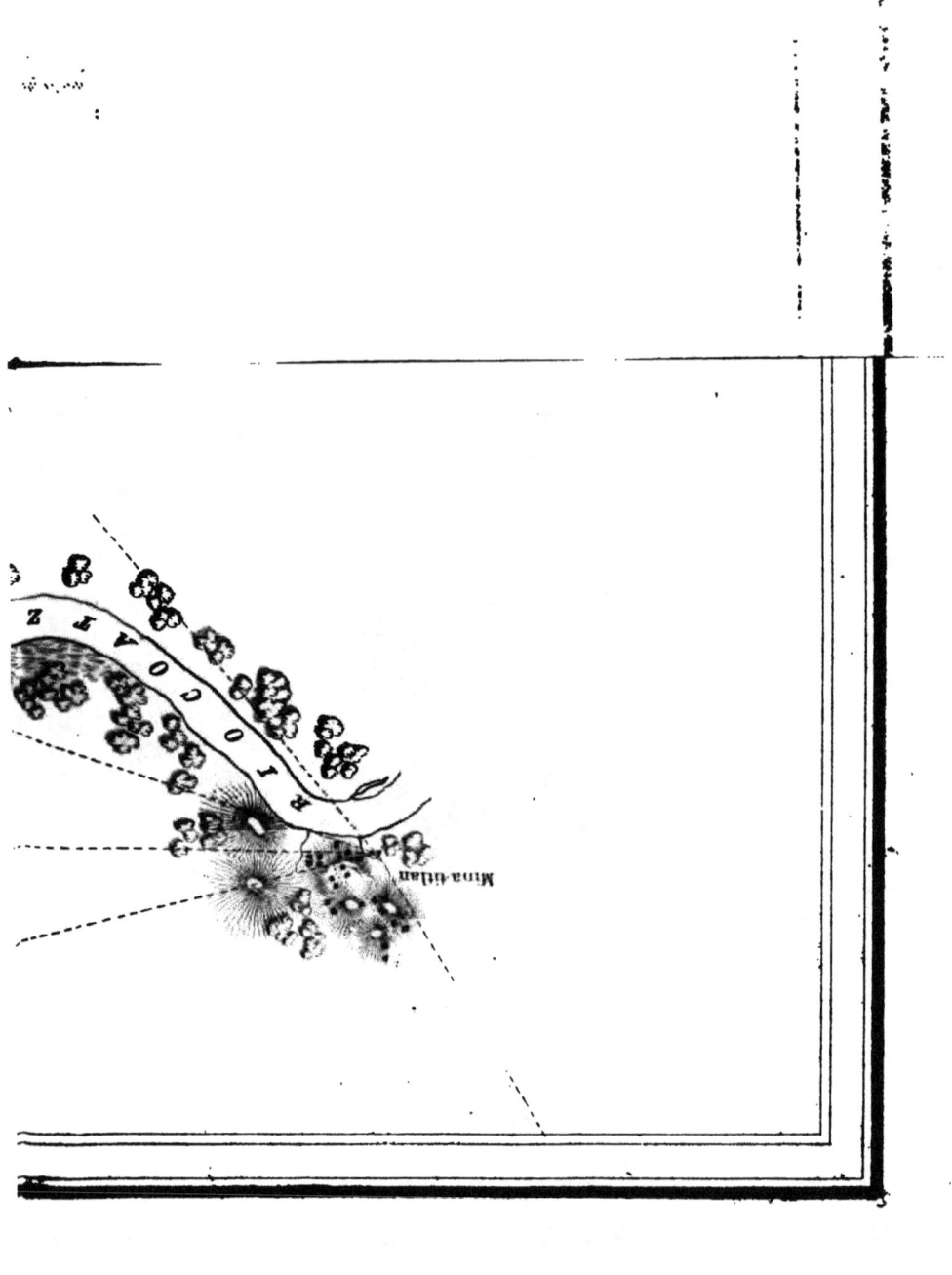

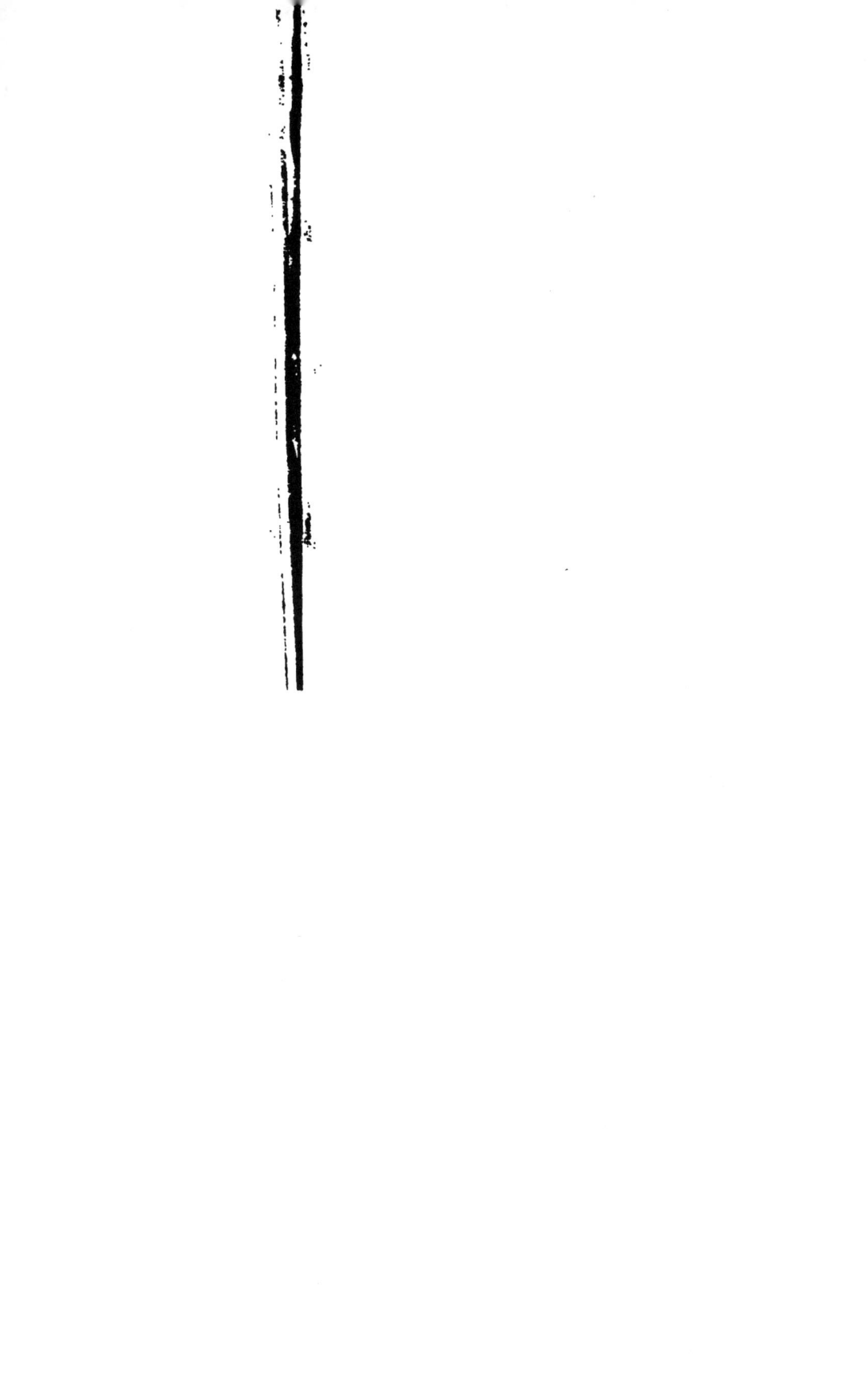

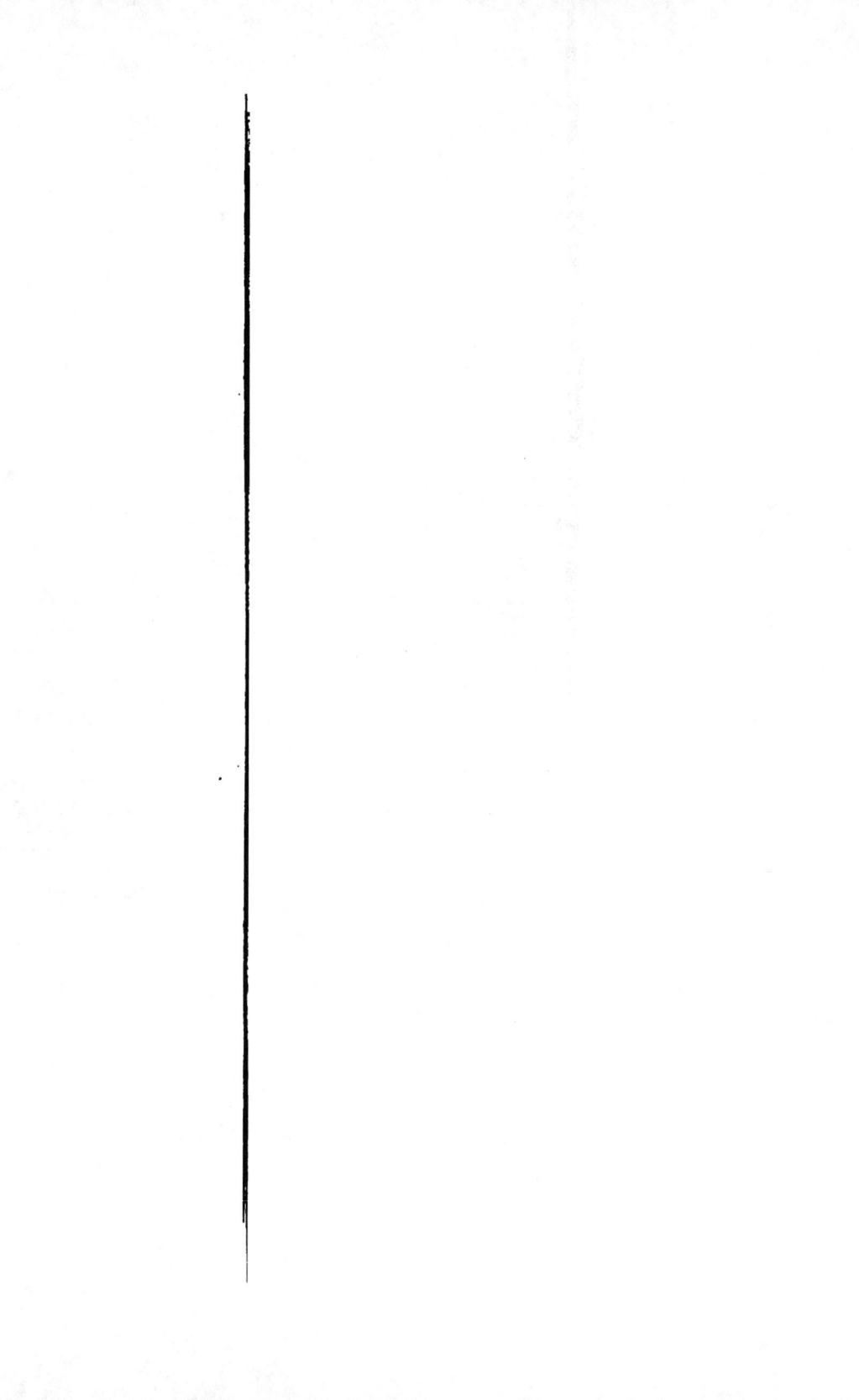

MAPS

ILLUSTRATING THE

ISTHMUS OF TEHUANTEPEC.

NEW YORK:

D. APPLETON & COMPANY, 200 BROADWAY.

M DCCC LII.

CONTENTS.

MAPS.

CPSIA information can be obtained
at www.ICGtesting.com
Printed in the USA
BVHW040049241118
533517BV00019B/362/P